Why not Top Bar Hives?

A look at the practicalities of keeping bees in Top Bar Hives

J R Slade

NORTHERN BEE BOOKS

Why not Top Bar Hives?

A look at the practicalities of keeping bees in Top Bar Hives

ISBN 978-1-908904-42-3

Published by Northern Bee Books, 2013
Scout Bottom Farm
Mytholmroyd
Hebden Bridge HX7 5JS (UK)

Design and Artwork, D&P Design and Print
Printed by Lightning Source UK

Why not Top Bar Hives?

A look at the practicalities of keeping bees in Top Bar Hives

J R Slade

Foreword

Keeping bees in top bar hives is becoming more popular with amateur beekeepers and even with some commercial bee farmers - the latter especially in the United States. For those who choose to have just a few colonies because they are interested in the fascinating world of bees and wish to produce enough honey for themselves and their friends, it is an easy and economical way of entering the craft. D-I-Y beekeepers will certainly enjoy constructing a hive, often cheaply from pieces of re-cycled wood, and this booklet will provide a useful guide from an author whose designs and innovations are based on experiments and observations over many years.

One of the most interesting and important points about this method of bee husbandry is that it allows the bees to build their nest in a natural way. Whilst the bees are given top bars on which to build their combs, they can construct combs in their preferred shape and with the right balance of worker and drone cells - something which is not possible in standard frames fitted with worker foundation. Drones are essential to the colony's well-being and the dearth of drones, due to the elimination of wild nests because of varroa, the constant use of worker foundation and the destruction of drones by beekeepers, has enormously diminished the gene pool. The resulting queen failure is a possible reason for the collapse of colonies. Allowing colonies to build all the drone comb they require may well help to redress this dreadful situation.

Another important factor is that the bees are allowed to build their combs without the use of any foundation. Bees need to build wax as part of their colony life and it is better for them to be allowed to do this rather than to draw out wax combs from foundation of dubious and possibly adulterated quality.

Following the methods described by the author, the beekeeper will be able to handle the combs safely without them collapsing and experience the thrill of seeing the bees working in the way nature intended.

As for the honey; it is good to be able to take a piece of comb now and then from the hive, as soon as it is sealed, and enjoy the different tastes which the flowers give to honey throughout the season.

John Phipps
Editor
The Beekeepers Quarterly
Greece
June 2013

Contents

Introduction

There is a lot of information on top bar hives of different types and unfortunately many people have been gulled into trying them without those supplying them offering any form of in-depth management assistance.

Top bar hives are the most wonderful way of keeping bees, as the bees are in a structure closest to their natural environment, and one in which they can be easily managed and observed.

Top bar hives are not for those beekeepers hell-bent on a vast honey crop, but for those who wish to have bees for the wonder of the bees themselves; the acquisition of a crop of honey is very much secondary.

Many lady beekeepers are put off beekeeping because of the lifting that is required with conventional hives and the variable defensiveness of the bees within. The reality is that other than the large (but not heavy) roof of the top bar hive, there is no heavy lifting and from experience the humour of bees in top bar hives is always so much better. The fact that bees in top bar hives produce their own comb without the provision of foundation I believe is a great factor in the contentment of the bees and their docility.

The object of this booklet (the first part of which is a slightly edited copy of an article published in the Beekeepers Quarterly - issue 98, November 2009) is to provide sufficient information so that anybody embarking on top bar beekeeping knows some of the things that work and some of the primary pit -falls. The second part is a suggestion for top bar hive design based upon experience in the use of TBHs from then till now (2013).

Part One

A Brief History

Hunter gatherers sought out wild nests of bees and robbed them of their honey. However, when people began to live settled lives and started to farm it was an enormous advantage to have colonies nesting nearby. Initially, hollowed-out logs were used to attract the colonies, but harvesting the honey often resulted in the loss of the colony, too. Over time, most likely beginning in Greece several centuries ago, beekeepers realised that if they placed wooden bars at a particular spacing on the top of a receptacle which had inwardly sloping sides, the combs would not be fixed to the hive walls, so the honey could be harvested without the destruction of the whole nest. The idea of the sloping sides is the main feature of top bar hives and, particularly in beekeeping development, 'long' top bar hives of this pattern have become commonplace.

From this point onwards, many beekeepers have looked at top bar hives and a whole range of designs have emerged, some more successful than others. The object of the development is not to disparage the designs of others but to look constructively at what has gone before and develop a more practical top bar hive.

Designing a starter Top Bar Hive

One of the first things that had to be determined was size. The size was set not completely arbitrarily, but as the hive was to have a single chamber a guesstimation was made that the chamber should have approximately the same volume as a National hive brood and half.

When bees construct comb in a free space the comb takes a catenary form, so to get the approximate volume that the bees might require it was decided to have 24 top bars each 450 mm (18 ins) long. All of the top bar designs that were looked at used top bars that were the same width as the frame spacing in conventional hives, ie 35mm (13/8 ins). With the top bars being 35mm they could be "crowded", so that there would be no space between them, thus obviating the need for a crown board.

There were also other design factors that had to be considered:

- Whether the sides should be vertical or angled so as to approximate the catenary shape of the comb.
- Position of the entrance.
- Size of the entrance.
- Whether or not there should be a varroa mesh screen with a removable examination tray.

- Provision for feeding.

From these simple questions the following were determined:

- From a simple experiment with a length of chain, an angle of 15 degrees was arrived at for the sides.
- Quite arbitrarily the position of the entrance was set at a ¼ of the length of the hive from the end of the hive and a ¼ way up the sloping side from the bottom.
- From experience a relatively small entrance was decided upon but with a twist. Some beekeepers say that an alighting board is beneficial. It was therefore decided to put in a double entrance (See photo). The overall size was to be 150 mm by 12 mm high.The entrance was partitioned and an alighting board applied to one half. The intention was to ascertain if the bees leaving and returning had a preference.
- Because the first top bar hive was wholly for experimentation it was decided not to have a mesh screen but to have a removable slide-out panel to check for varroa.
- Because the top bar hive did not require a crown board, it was decided that an internal "frame type" feeder be constructed.

1. Double entrance arrangement in the first top bar hive.

On completion of the top bar hive a swarm was introduced and feed applied by means of the frame feeder. The top bars had previously been fitted with starter strips of foundation to encourage the bees to put the comb on them so that the top bars and drops of comb could be removed and examined.

Observation of the first starter Top Bar Hive to ascertain some of the problems

The most striking thing on looking into the hive after the first week was the extent to which the bees had built comb. Expecting similar results to those observed under similar circumstances in a National hive, it would not be an exaggeration to say that possibly twice as much wax had been produced. I had been used to seeing wild comb in all sorts of configurations, but the bees had in this instance stuck to the plan and the comb was perfectly aligned onto the top bars via the starter strips.

It was at this point (first manipulation) that certain aspects of top bar hives presented themselves as being possibly problematic. Having found that the bees had produced copious amounts of comb, it was noticed that they had done something most unexpected. For the bees, the top bars constituted a single solid roof; they appeared to want longitudinal chambers immediately under the top bars. This can be seen clearly in the photograph below and I have also drawn a line diagram to illustrate this.

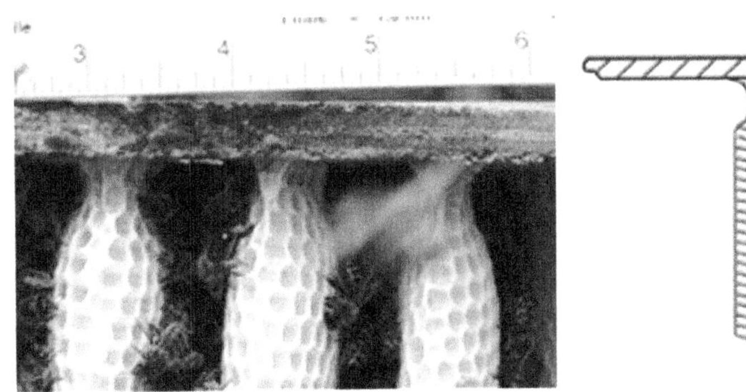

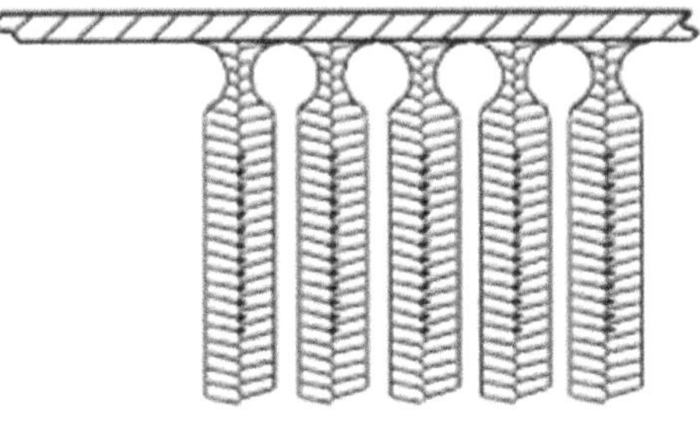

2. Diagram of comb thinning.

This thinning at the top of the comb must be of use to the bees but for what purpose? My feeling is that they require air movement and these are perhaps fanning channels, so that they can draw air up and out across and along the channels. Whatever the reason the narrowing of the comb made manipulation very difficult. It was possible to lift the comb straight up on the top bar and examine one side but rotating or reversing the comb to examine the other side proved almost impossible.

The second thing that became immediately noticeable was when closing up the top bars. The bars have to touch along their full length; getting the bees out of the way was difficult. When smoke was used some bees went down as required, but a great number remained on top of the bars which gave them no route back. The only thing that could be done was to use more smoke together with a brush to clear the top.

I mention the use of smoke to clear the bees from the bars so that the top bars could be closed up, but this was really the only time that smoke was required. The temperament of the bees in the top bar hive was quite remarkable. The normal aspects of manipulation applied in that the bees were inclined to get everywhere but there was no aggression and the smoke was used merely to move bees out of the way as stated when closing up.

The double entrance proved to be very interesting in many ways. My thoughts when creating the entrance was that bees leaving the hive might go via the side without the alighting board and the returning bees would use the side with the alighting board to land. This assumption proved to be almost totally false. From close observation the bees preferred to fly straight into the hole whereas bees leaving the hive clearly preferred to crawl out onto the alighting board and fly off from there. This phenomenon became even more pronounced the busier the hive became, to a point that when activity was at its greatest no bees were landing on the alighting board but flew straight into the hive via the hole. At the same time the alighting board was exclusively used by bees leaving the hive. The fact that the returning bees preferred to fly straight into the hive led to other experiments. (See footnote 2)

A further observation was the way in which the bees supported the comb. Having produced a comb that was narrowed at the top,the bees then braced it at its edges to the sloping wall of the hive. What appeared at the outset as a possible obstacle, proved not to be the case. They did brace - but only in the form of one or two "stitches" at each edge and invariably at no more than 100 mm down from the top bar. To remove a top bar with its attached comb a simple "parting job" was required with a hive tool; once done the comb lifted straight out.

Changes to the design based on findings from the first Top Bar Hive

The size of the first top bar hive, whilst not having been scientifically determined was a guesstimation based on a National brood and a half. When designing the second hive the volume was reduced a little, not because the first was considered too large but because observations from the first required us to take a different course.

The prime factor was the top bars themselves. The problems of closing up during and after manipulation suggested that a narrower top bar should be used. The decision was made to use the top bars from standard frames with standard spacers. The hive then became slightly narrower. The depth remained approximately the same.

The entrance on the first top bar hive was along its length, on the second hive it was decided to put it at the end. The thinking behind this was that manipulation is normally carried out with the hive entrance facing away from the beekeeper. With the first hive this entailed reaching across when lifting a bar and comb. By having the entrance at the end manipulation would be as for a standard hive with the frames the "warm way"

To overcome the problem of handling the bars with the attached comb, a "peg system" was devised using standard frames. After removing the clamp bar from a standard frame top rail, a 6mm hole was drilled vertically through the thick section of the top bar near the centre length-wise, so that the edge of the hole coincided as closely as possible with the inside edge where the clamp bar fitted. By whittling the end of one of the redundant bottom rails of the standard frame to about 6mm it could be forced into the drilled hole and fixed with a gimp pin. The "peg" was cut down in length to about ¾ of the depth of the hive. (See below) By using standard top rails from standard frames there was also the facility to put in a starter strip of wax using the clamp strip.

Drilled hole in standard top rail.

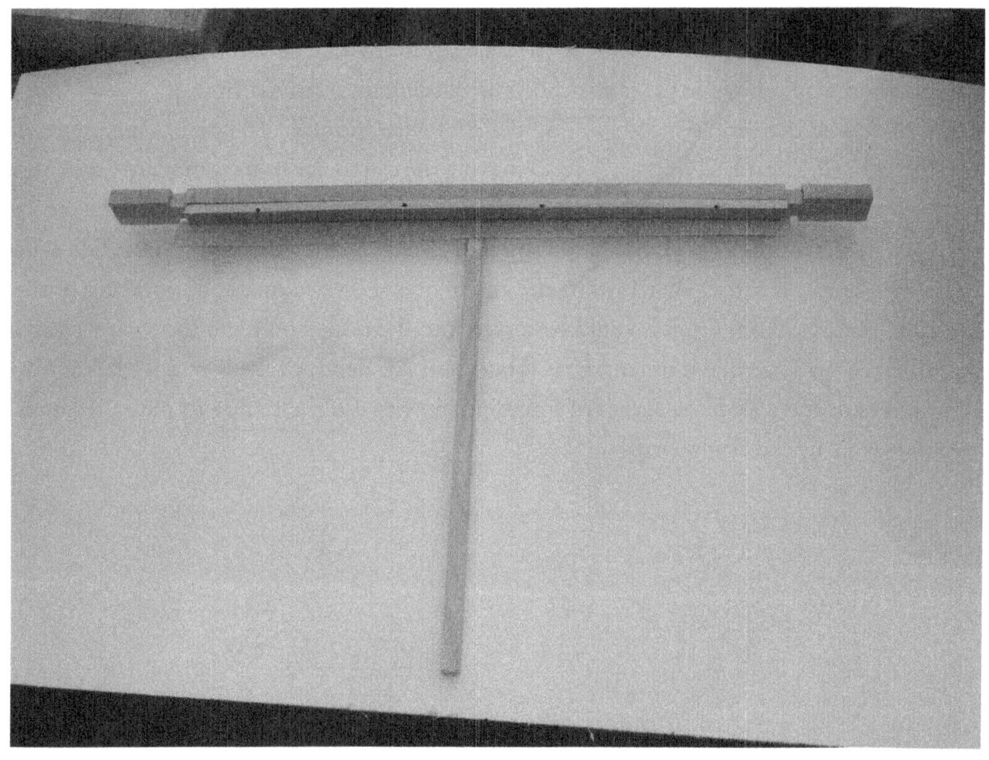

4. Peg in place with strip of foundation.

The decision to use standard-dimensioned frame tops as top bars meant that a crown board would be required. This then allowed for top feeding through a Porter bee escape hole.

Rather than having a single large crown board it was decided to have two half

sized ones (When top feeding with National hives space can be created for a feeder by using an empty super as an eke).

For the new top bar hive it became necessary to manufacture and use an eke. As with the first top bar hive, a frame-type feeder was also made for rapid high volume feeding.

The new top bar hive was fitted out with the standard frame top, top bars with their "pegs".

When a swarm was placed in the hive the reaction was very much the same as with the first top bar hive; the bees responded equally dynamically and built comb in a matter of days. There were, however, fundamental differences in the way they did so. With the original wide top bar the bees started with a single segment of comb which they enlarged and enlarged until it fitted the section of the hive. With the narrower top bars with the pegs they started with two segments, one either side of the peg like tear drops. As they extended the comb down, the comb took on the appearance of a "w". The further down the comb extended the smaller the centre peak in the "w". Once the comb had reached the end of the peg it was finished in exactly the same way as in the first hive. When the comb was complete the peg disappeared within the comb. Another thing that came to light when using the narrow top bars was that the bees did not produce the narrowing at the top of the comb that had been observed with the wide top bars. *(See footnote 1)*

5. Drop of comb from the first top bar hive with "peg" fully encased. NB, especially on the left hand side near the top, there remains the remnants of a "stitch" where the comb was braced, whereas the whole bottom 2/3rds was finished without any bracing to the walls of the hive.

The "peg" system proved to be highly successful to the point that the top bars and comb could be manipulated as easily and safely as any normal deep frame from a conventional hive. Faced with the reality that the first hive's top bars were almost impossible to manipulate, it was decided to retro-fit "pegs" into the first hive. To achieve this 6.5mm holes were drilled as vertically as possible in the centre of the top bars. So as not to cause too much upset to the bees the drilling was carried out using a small wheel brace. The holes hopefully would be on the foundation line ie at the centre of the comb. Once the holes had been drilled pegs in the form of 6mm wooden knitting needles were pushed down through the holes into the comb below. The needles (pegs) were fixed in place with small wood screws, screwed in on the edge of the drilled hole.

6. Pushing the needles through the top bars into the comb.

The hive was not disturbed for 7 days, thus allowing the bees time to repair any damage caused when inserting the pegs, and also to give them time to fix the comb to the pegs.

On opening the hive a week later the results were better than expected. One or two of the pegs had not gone straight down the centre of the comb but had veered. This was almost certainly due to the fact that the drilling was not accurate enough and not all of the holes were vertical. None the less, even where the pegs had veered the bees had attached buttresses to the pegs. With the pegs in place it immediately became possible to manipulate the bars and comb as easily as those in the second hive.

7. Section through the second top bar hive.

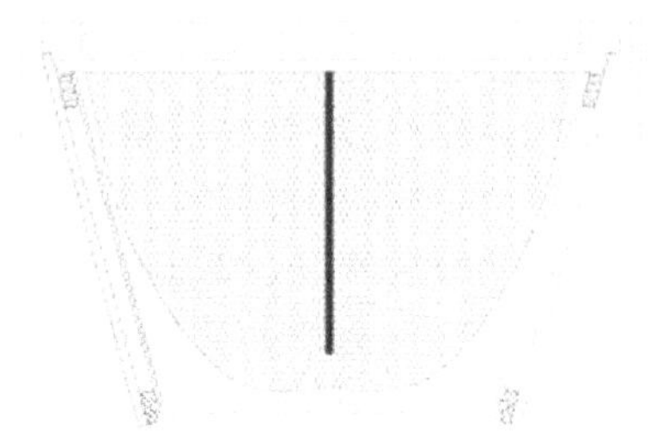

Comb with peg in place.

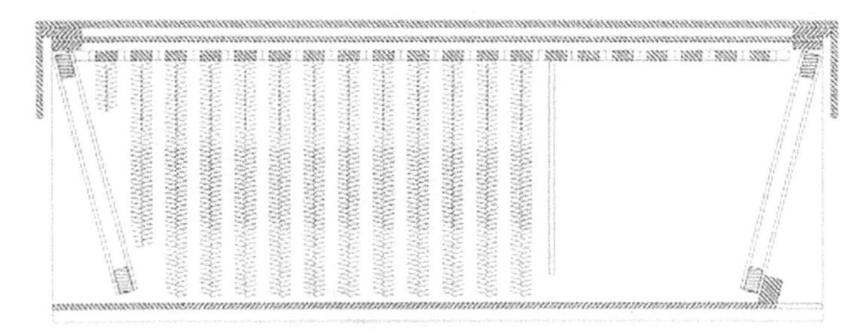

8. The entrance is below the left hand angled-end.

The two end walls of the hive are identical and as can be seen from the diagram they do not reach down to the floor. The floor is in two sections, a shorter section acts as an alighting board and the bees pass under the end wall. The other longer section acts as a removable panel for checking varroa drop. The longer section has a stop block fitted that closes off the gap under the end wall. It can also be seen from the diagram that the hive is constructed with double skin walls. This means that the inner wall always remains dry.

Manipulation of the first Top Bar Hive

9. A pegged top bar with fully-formed comb being manipulated in the same way as "conventional frames" would be.

From the observations of the first hive it was also decided not to have a varroa mesh but retain the slide-out panel for checking for varroa. The reason for this was simple. We had already questioned ourselves as to the efficacy of mesh floors in terms of pollen loss. We had noted from the first hive that whenever we checked the slide-out panel, whilst finding varroa, we encountered little or no pollen. As to whether the varroa count was accurate without a mesh, we felt the retention of pollen by the bees was more important. Another surprise was that despite the fact that the entrance was several inches above the slide-out panel the bees kept it very clean with only the smallest amount of debris present at any one time. What we found does beg the question; are varroa mesh screens necessary? *(See footnote 3)*

From the outset the object of the top bar hive development was to look at top bar hives with the older or lady beekeeper in mind. Not every beekeeper old or young is on a mission to be the beekeeper to produce the greatest crop of honey. On the contrary a large number of people want to keep bees for the pleasure they get from one of the most enchanting occupations in the world; others for the benefit that honeybees bring to the environment. Whatever the reason an individual might have for keeping bees there is a place for the top bar hive.

From the use of the two top bar hives we found a number of things that are quite significant; one being that the lady beekeepers that have joined us in the development do like them very much. There are many reasons but one is the temper of the bees, which when compared with other hives are noticeably less defensive.

The development so far has identified some of the problems or idiosyncrasies of top bar hives but there is some way to go yet.

Footnote 1

Whatever the bees created the "chambers" for in the first hive they did not require them in the second hive. Do the bees need either air passages or bee passages at the top of the comb, or above the comb but under the crown board? I conjecture from this that the bees do require an internal circulatory ventilation system.

When they occupy a void such as a hollow tree they have perhaps evolved a system of chamber construction that permits air to be moved readily around and through the cluster. Since the cluster or main body of bees would be warmer than the air surrounding it, then possibly the air rising through the cluster is then drawn laterally and down the side walls. In warm weather their ability to remove heat would be much enhanced; however in the winter these chambers would be filled with bees and all convection currents stopped.

A clue to this might be that in very old combs in a hive, small holes can frequently be found. Whether this is for air circulation or simply for short cuts for bees to get quickly to another part of the hive, I am not sure.

If this conjecture has any merit then this form of ventilation would also be a very efficient way of removing moisture from the hive. Air that is drawn up through the cluster would be warmed and have moisture added. As it descends against the cooler wall of the hive, some moisture would be lost. An internal ventilation system would be infinitely better than an unregulated through system as the bees would be able to control the temperature difference exactly. This leads to another question with narrow top bars; do they provide an acceptable artificial chamber at the top of the comb? If so, should the crown board have a bee space on its underside or should the crown board sit immediately onto the top bars?

Footnote 2

Having observed the way that the bees used the double entrance we decided to carry out a little experiment on what type of entrance bees preferred. We used one of my own varroa floors which had a full entrance width, and a mesh screen fixed to its upper surface, so that there was just a bee space between the bottom of the frames and the mesh, and a ramp over the first 40mm (1½ in) of mesh.

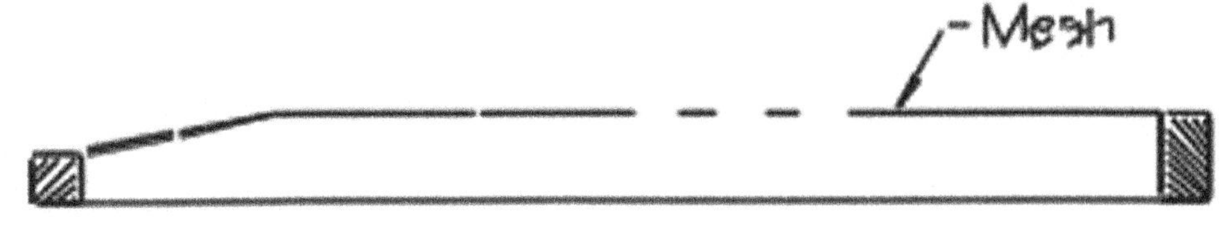

10. Section through mesh floor showing the first 40mm of mesh angled to form a ramp.

On one side of the entrance a piece of the side wall of the floor was cut out so that an entrance block could be slid in from the side *(This can be seen on the photograph below)*. On completion of the modification to the floor an average- sized colony was chosen and its floor replaced with the modified one.

Firstly the entrance was left full width. The movement of the bees in and out appeared disjointed in that there was bee activity in the entrance but the bees comings and goings seemed to be uncoordinated. Bees from inside milled around the entrance without leaving the hive. After a few days the entrance was set so that approximately 50% was closed off. With an opening of 22mm by 200mm the random activity of the bees inside the entrance was much reduced.

Over a period of about a month the entrance was reduced in stages so that finally the entrance was near square (22mm by 22mm). As the entrance neared its smallest size, two things became quite apparent. Firstly, the bees clustered around the opening on the hive front and, secondly, there was what might be described as a swarm effect with a large number of bees queuing in the air between 1 and 2 metres from the hive. As the entrance was enlarged the swarm effect diminished as did the number of bees on the hive front. When the entrance size reached approximately 22mm by 100mm the swarm effect vanished as did the bees on the front. What further became apparent was the way in which the bees came and went.

Almost like magic there were no bees in the entrance, rather the bees exited on the wing from within the hive and returning bees flew straight into the entrance. At this opening size the bees appeared to come and go in what seemed to be an efficient way. Several days after making observations and setting the entrance to what I considered to be the ideal, I happened to be in the apiary in very inclement weather. Looking at the hive I noticed that the hive entrance had been reduced to a mere finger sized hole by what might be described as a wall of bees. With the opening set as it was and with the weather being so cold and wet, could it be that the entrance was of a manageable size and the bees were operating some sort of "variable ventilation valve"?

One further factor when looking at entrances that may be of relevance is security! The first top bar hive has an entrance leading directly onto the side of the comb. The bees made no modifications to the comb near or at the entrance. From observations as the autumn drew in it was noted that wasps appeared not to attempt to enter the hive. Whether this is due to the fact that bees occupied comb immediately inside the entrance or for some other reason is not clear. Whatever the case may be, I think that further investigation into an ideal entrance size, position and configuration should be carried out.

11. The wall of bees as described above.

Footnote 3

The fact that little or no pollen was found on the slide-out panel, suggests that pollen dropped by bees was either collected up or cleaned away as other debris. Since pollen is of such importance to the bees it seems more likely that they collect and use it. Asking other beekeepers, I have yet to find one who has seen bees discarding pollen.

Part 2

Thoughts on the final design

The use of top bars that are 17" (432mm) long x 7/8"(22mm) wide x 3/8" (9.3mm) thick, ie the same section as a standard frame where the spacer fits, proved to be very successful, thus determining the width of the hive.

Again the angled sides proved to have two primary advantages; firstly as the drops of comb have no "frame ends" the bees were far less inclined to attach comb to the walls of the hive, secondly rain water running off the roof does not run down the outside of the hive and the wall remains dry.

The ends of the hive on the original design were angled in at 15 degrees the same as the walls, this proved a bit of a nonsense, and the bees appear to prefer a full sheet of comb at the end.

The only thing left regarding the main box was the length. The original hive was designed to take 30 top bars but this proved to be excessive and from observations the hive need only be long enough to accommodate 18 top bars on narrow spacing (18 x 37.5mm) = 675mm plus 20mm as a manipulation space - which gives an internal length of 695mm (27 3/8").

The fixing of a nominal size for the box was the easy bit. What is lacking in terms of top bar hives is a coherent management system centered around a design. In this respect certain questions have to be addressed:

1. **Is there a need for a varroa mesh screen?** Absolutely not! Varroa screens in conventional hive may be de riguer but the loss of pollen through them is great and that loss is not compensated for by additional knowledge regarding varroa.

2. **Should the bottom of the hive be open save for a mesh screen?** No; from all I have gleaned the bees are profoundly affected by cold and draughts.

3. **Where should the entrance be?** After playing about with entrances of differing sizes and location, I have come to believe that end entrances are not liked by the bees, nor are bottom entrances. The best location is approximately 1/3 of the way up from the bottom and about 8" (200mm) from the end. *(See photos)*

4. **How big should the entrance be?** The ideal entrance size is much smaller than one might expect, and within reason, the smaller the better. The entrances in the hives in the picture represent the smallest being approximately 2½" x ½" (65mm x 13mm) The maximum size should not exceed 4" x 1" (100mm x 25mm)

5. **How should spring, summer or autumn food be supplied?** There are a number of feeders available but none are suitable for TBHs. The solution was to have drop in feeders that utilize a 1 litre feeder bottle or two 1 litre feeder bottles. Whether it be after housing a swarm, or for emergency feeding, a feeder is essential.

6. **How should winter food in the form of fondant be supplied?** Winter feeding is sometimes needed especially if we have an extended winter. Going below the crown board is a no no, therefore the only way is with a form of top feeder that can be used with fondant. *(Refer to 7 below)*

7. **Should there be ventilation, summer and or winter?** Ventilation is always a topic of much debate. If the bees are inactive as is the case through winter then even the smallest entrance will provide more than adequate movement of air. If the bees are active then they will self-ventilate through what ever entrance they have. One type of ventilation that is essential is that needed to keep the upper part of the hive dry. An eke approximately 2" (50mm) deep that has vents, sitting on the crown board, provides a ventilated space above it. This space can be utilised to accommodate a fondant feeder over one of the holes in the crown boards that are normally closed using a piece of plywood. In extremes of very hot weather then a hole/holes in the crown board can be uncovered and excess heat can dissipate through the vented eke.

8. **Is there a nominal over-winter number of drops of comb?** At the end of the bee season the bee will tend to cover a certain proportion of the drops of comb, any totally empty drops or partial drops should be removed and a dummy board put in. If there appear to be insufficient stores then having a segmented crown board system will permit feeding without removing the crown board immediately over the cluster.

9. **How should the TBH be kept clean?** Unfortunately TBHs can not exist singularly; however many TBHs you have you must have a spare. Every year (or at least every other year) two things must happen; one, old comb has to be replaced, two, all drops of comb must be transferred to a clean hive and the old hive cleaned up.

12. Top bar hive with 9 top bars and feeder ready to receive a swarm.

The hives shown in the photographs have four entrances, these were put in for experimental purposes but in practice it was found that a single entrance was sufficient.

13. Hive with three-section crown board with plywood panels over the holes.

14. Hive with centre crown board removed to expose feeder.

15. Eke in place on the crown boards. There are two vent holes in the eke, this allows the free movement of air over the crown boards. The vent in the far wall on the right is clearly seen, the other is diametrically opposite.

16. Drop in feeder with 1 litre feeder bottle.

17. Drop in feeder with 2 x 1 litre feeder bottles on adaptor.

18. Top Bar Hive complete with roof.

NB - the roof is deep enough so that it extends downwards to prevent rain etc from reaching the crown boards. The drip line from the roof is also outside the small alighting/ closure block. The size of the roof makes it a little unwieldy therefore having two handles makes removal and replacement much easier. Removal and replacement of the roof is also made easier by it being generously larger than the top of the hive and eke.

19. Bees entering and exiting the hive.

20. Hive on a stand with a securing rope.

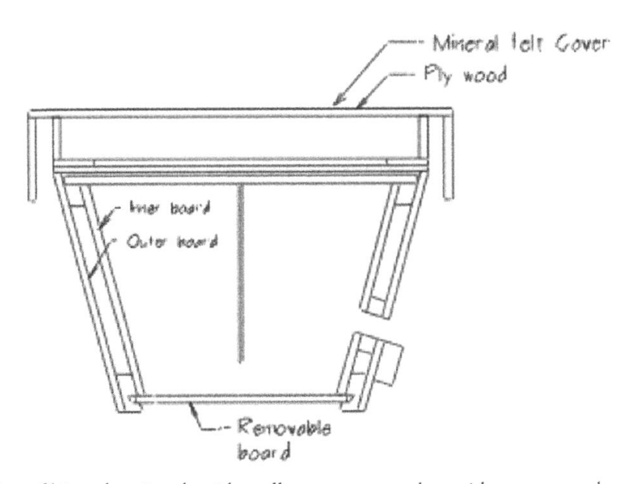

21. Cross section of hive; showing the side walls, entrance, top bar with peg, crown board, eke and roof.

The side walls are double-skinned for warmth and moisture protection. The outer-board of the side can be extended as above so that a flat crown board can be used and provide a bee space between the top bars and the underside of the crown board. If the outer-board only extends to the same height as the top of the top bars then a crown board with a bee space on the underside would be required. The removable board groove has to be wide enough so that the board will slide easily.

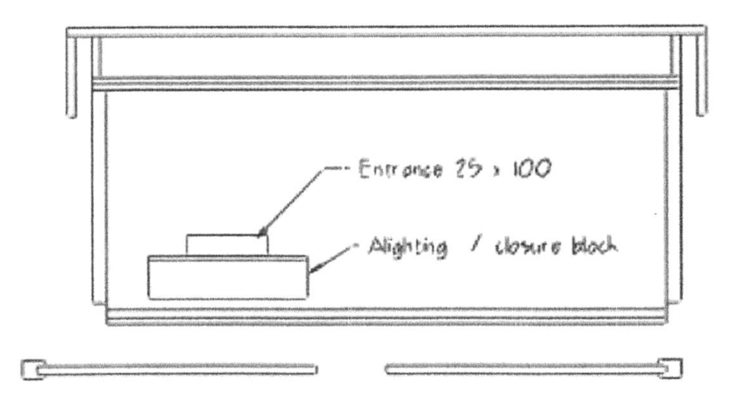

22. Front view semi-section of the hive showing entrance alighting/closure block.
Removeable board shown out of place.

Although not shown, the end walls of the hive are double-skinned - again for warmth and dryness. The end walls extend down to approximately 10mm above the upper edge of the removeable board groove. This is so that as the board is drawn out, any debris on it is not scraped off. The removeable board has beads serving two purposes; firstly so that there is a step for fingers under the end so that the board can be drawn out and, secondly, to close the 10mm gap.

23 - 30. The following photographs show the manipulation of drops of comb. It can be clearly seen that by using top bars that are 7/8" (22mm) x 3/8" (9.3mm) with standard narrow spacers and a central peg, the drops of comb are easily and safely handled. What cannot be seen in the photographs is the quietness of the bees and the reality that no smoke was required at any time during the examination of the hive, other than "to move" bees, when closing the hive so as to prevent crushing of bees.

23.

24.

25.

26.

27.

28.

29.

30.

Feeders

For feeding syrup in the drop-in feeder I use square section, rigid-plastic bottles purchased with fruit in. They are available from most supermarkets with a range of fruits. They are approximately 90mm x 90mm x 155mm tall. To convert one to a feeder you also require a plastic flour sieve. Again the sieves are available from most supermarkets. One sieve will provide enough mesh to make five or six feeders.

- There are a number of tools and operations needed to make the feeder.
- Tools:
- Some form of hole cutter, a pair of scissors, a hot glue gun, circular weight (which must fit the inside of the lid), and a roll of cling film.

What to do:

1. Cut a hole about 45mm diameter in the lid of the bottle. The hole does not have to be round; a square hole cut using a Stanley knife is just as good.

2. Using a pair of scissors cut the mesh off the frame/handle of the sieve as closely as possible to the frame.

3. Cut the mesh into as many pieces as possible so that each piece will cover the hole in the lid with about 10mm over-size all round.

4. While using the hot glue gun the glue must be as hot as possible. When fully heated, place a ring of hot glue around the hole on the inside of the lid.

5. Whilst the glue is still molten lay on the mesh.

6. Place a small sheet of cling film over the mesh.

7. Place the weight on the cling film to compress the mesh into the molten glue.

8. When cooled remove the weight and peel off the cling film. The cling film is used so that the glue that will pass through the mesh does not stick to the weight.

The result can be seen in the following photos:

31. Feeder bottle with lid on.

32. Feeder bottle and lid showing the glue and mesh on the inside.

For fondant feeding I use shallow, rectangular food storage containers which, again, are available from supermarkets. The size I use is approximately 160mm x 125mm x 40mm deep.

The only operation needed to convert one to a fondant feeder is to cut a hole, approximately 40mm diameter, in the centre of the lid.

33. Container with lid removed.

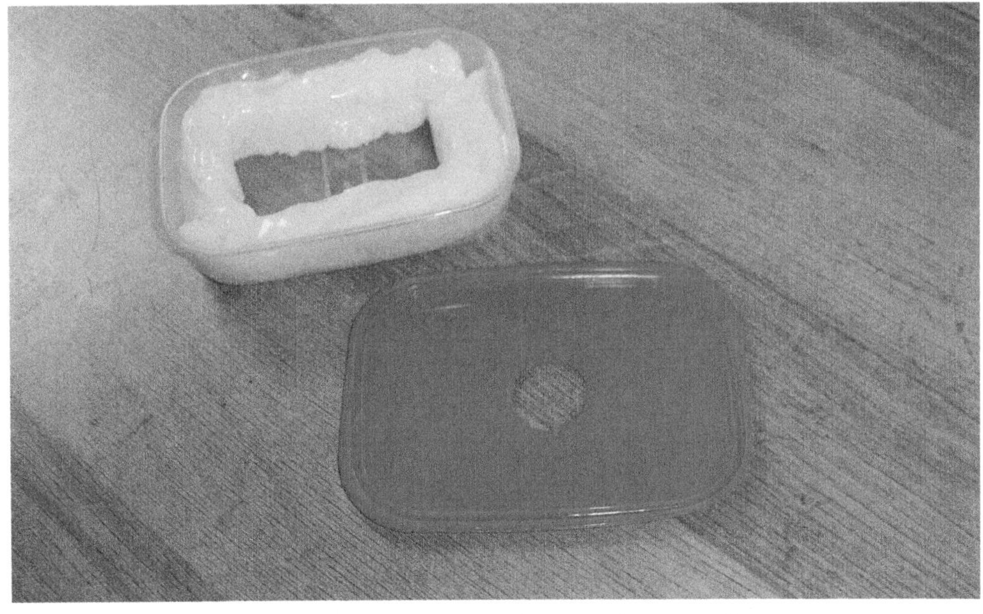

34. Container with ½ kg of fondant pressed into the walls.

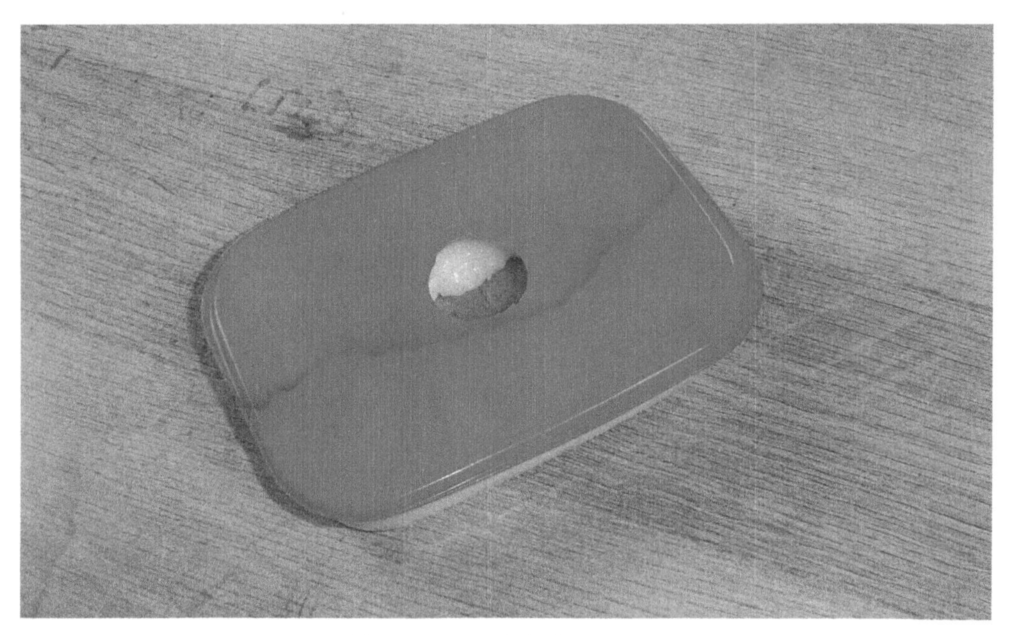

35. Container with the lid replaced.

36. Feeder inverted as it would be placed over the hole in the crown board. Placing the fondant around the walls of the container permits the observation of bee activity.

37. Fondant feeder on the hive, with bees taking the fondant down.

Tools

In addition to the obvious requirements for beekeeping such as a bee suit, smoker, standard hive tool etc, one other tool is essential, and that is a long flat implement that can be used to break any comb attachments that the bees make to the wall of the hive. Although with the angled side bees tend not to make as many attachments as with vertical sided hives they do, however, make some. Whether in conventional hives, National or otherwise, there are always some odd bits of comb that the bees place on the walls of the hive. With top bar hives there is no wooden frame around the comb and so it becomes a must that any comb, or remnants of comb, after being detached have to be removed. The best tool that I have come up with is a standard good quality 12" stainless steel rule. The fact that the top bar hive has a removeable bottom board means that any debris created by scraping down the wall of the hive can be removed by pulling the board out.

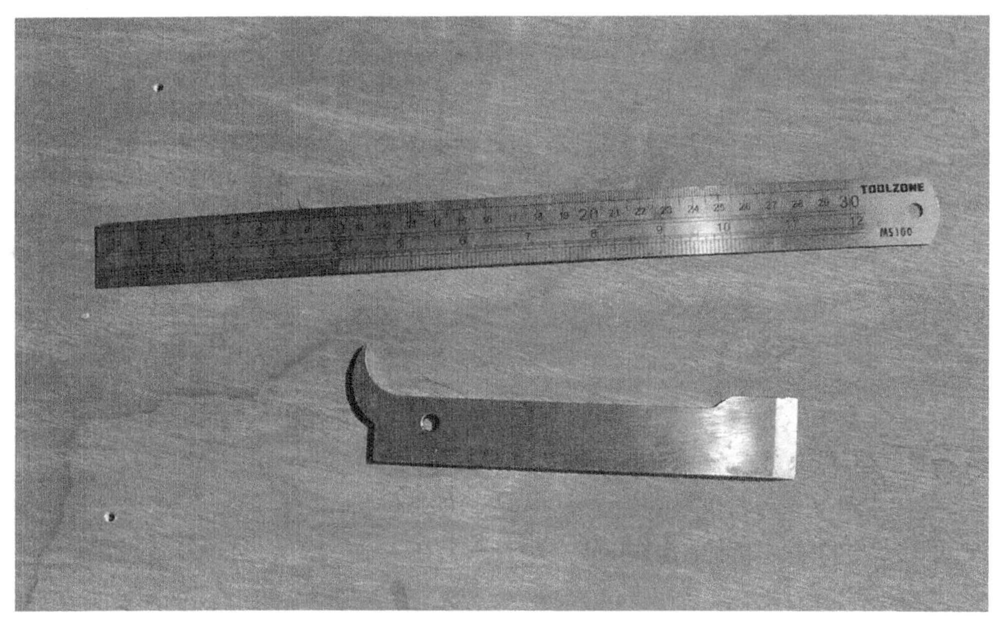

38. 12" stainless steel rule with small, handier, standard hive tool.

Stands for Top Bar Hives

Stands for bee hives have always been a matter of some debate; there are those that have to have proper, solid, permanent brick and mortar stands and then after that anything from a pile of concrete blocks to old milk crates.

One point of TBH beekeeping is the avoidance of lifting and anything else that impinges on the back. Therefore, a stand that places the hive at a level that does not require too great a stoop when manipulating the combs has to be the order of the day. I have tried a number of different stands and have come to the conclusion that the nearest to ideal is a four post, two top-rail stand. The whole stand is made from 2" x 2" and 2" x 1" graded, treated, timber and nails.

Cutting list for a two TBH stand
From 2" x 2" timber:

- Four posts 34" (860mm) long cut and sharpened to a point so they can be driven into the ground.
- Two cross rails 18" (460mm) long.
- Two long rails approximately 100" (2500mm) long.
- From 2" x 1" timber:
- Four upper cross rails 18" (460mm).

Construction

- Firstly, find a sheltered spot so that the stand can be placed with the long side facing south-east or south.
- The four posts are driven into the ground as a rectangle so the outside dimension front to back is 18" and the distance between the other two posts is about 60" (1500mm). It does not matter that the dimension may not be exact as the pairs of posts can be drawn together or opened a little so that the two cross rails fit. The important thing is that the tops of each post are level. To get them level drive all four posts in, then using a spirit level and one of the long rails to determine which is the lowest. Having done that; drive the other three posts into the ground one at a time, to the same approximate level.
- To fix the cross rails to the posts you will need 4 x 5" (125mm) long galvanized nails. To prevent the cross rails from splitting, drill holes in the centre and 1" in from the ends. The holes need to be the same diametre as the nails. Having fixed the cross rails, place the long rails on as can be seen from the photograph and fix with 4" nails. Again drill pilot holes in both the cross rail and long rails. The pilot holes need to be approximately 2/3 the diameter of the nails. For example if the nails are 5mm diameter drill pilot holes about 3 - 3.5mm diameter.
- The final part of the construction is the fixing of the four upper cross rails. They are placed about 122" (300mm) either side of the posts and cross rails and fixed with 2 ½ or 3" nails, again drilling pilot holes as before.
- Lastly, check the level. If the posts have gone into the ground as a result of fixing the rails, correct by striking the top long rail immediately over the posts. The whole construction can be seen in the photograph.

NB. When nailing pieces of timber together, always drive the nails in at an angle - the next nail angled in the oppposite direction. This holds the pieces of wood together more firmly making them difficult to pull apart. Also, to prevent the risk of splitting wood when using nails, tap the point of the nail with a hammer before hitting it in.

Posts that are sunk in the ground can survive for longer if charred with a blow torch - ensure that the carbonised wood is a few inches above ground level.

39. A stand with two top bar hives.

One small problem that can arise with this type of stand is distortion in the top rails. It can be seen from the photograph above that the left hand end of the front rail sags. To ensure that the hive remains level the use of wedges may be necessary. Another reason to have hives raised on stands is the reality that badgers find it near to impossible to upset them. With the addition of a rope fixed to the frame at the front of the hive and secured at the back, hives are also safe in windy conditions. The ropes can be seen securing hives in one or two of the photographs earlier in the booklet.

Where to from here?

From now on I do not think that any top bar hive will be built that does not have "pegged" top bars; as to whether full width top bars remain the order of the day or whether narrow bars with spacers will prevail I do not know.

The fundamentals from our development will form the backbone of future designs but other aspects of using this type of hive will have to be further developed. To this end a third hive is at present being designed that will look at the need for queen excluders, the creation of nuclei and swarm control.

With the information that will continue to come from the two top bar hives that we already have, together with fresh information that will come from the new design, we hope that by this time next year we will be in a position to formulate a definitive design that will provide a way of beekeeping that is not only practical for the beekeeper but just as importantly, for the contentment of the bees.

Part Three

(Update - Summer 2013)

Honey from Top Bar Hives

Taking honey from top bar hives is obviously not as straight forward as with conventional frame hives and as mentioned earlier in the booklet top bar hives should not be considered by those that are looking mainly for a large "honey crop"

One factor regarding the removal of honey from any hive is the use of queen excluders. Many beekeepers that use conventional hives do not use queen excluders. Those beekeepers accept the reality that some of the honey that they extract will come from comb that had brood earlier in the season and that the bees had cleaned the cells ready for the storage of honey. Many other beekeepers are of the mind that honey should only be extracted from comb that has never been used by the bees for brood rearing.

Quite obviously if a beekeeper wants to take honey in the form of cut comb then that comb must not have had brood in it and must also be pristine current season comb.

The same rules apply to top bar hives in that, if the beekeeper wants to have cut comb then he/she must select areas of comb that are new, (current season) and it must not have had brood in it. Unlike conventional frames the drops of comb from a top bar hive cannot be placed in an extractor. The normal way of taking honey from a top bar hive is to remove some of the drops of comb that have capped honey, then selecting areas that are suitable for cut honey and after cutting out sections with a comb cutter, cutting off remaining capped comb with a knife into an appropriate container, after which it can be pressed and filtered. Comb that has been "hacked about" can be placed back into the hive to be cleaned. This, as with many other beekeeping procedures has "I like to do it my way" elements. For this reason I shall not go any further as I know that each and every top bar beekeeper will develop methods that suit him or her best.

40. Comb cutter.

Swarm Control

Top bar hives that have narrow top bars with spaces and crown boards as described earlier in the booklet can be regularly inspected for the presence of queen cells and cups in exactly the same manner as for conventional frame hives. One of the problems with conventional frame hives when it comes to inspecting for queen cells, is the fact that bees very often remove parts of the comb down the sides against the frame and also along the bottom. The queen cells are concealed in these spaces and covered with bees. The finding of such cells can sometimes be very much hit and miss. With top bar drops of comb there is not a frame and therefore a gap between the frame and comb cannot be created; queen cells along the edge are readily spotted. Because there is not a frame around the comb the bees tend not to make holes through the comb. I have noticed that the bees do not fully embed the support pegs into the comb. On one side or the other the peg will remain exposed and the bees appear to like placing queen cells/cups on the exposed peg and again these are relatively easy to spot.

The fact that top bar hives can be inspected means that there is no need to dwell on swarm control beyond "cell finding" here as there are many other sources of information on the subject.

Creating additional colonies by means of nuclei

The spotting of queen cells is an integral part of "natural nucleus" production. Swarm control in its simplest form, is the identification and removal/ destroying of queen cells or the removal of the old queen and the removal/destroying of all but one of the queen cells. The remaining queen cell goes on to provide a replacement queen.

Creation of nucs follows a similar line; the only significant difference is that rather than destroying unwanted queen cells, drops of comb with queen cell/s attached are transferred to another hive/s (nuc boxes). Normally 3 or 4 drops are transferred to create a nuc. The number taken to create a nuc depends on the size and vigour of the donor hive. In some instances it may be possible to create more than one nuc from a very strong hive.

To create a nuc there must be available a suitable nuc box. With conventional hives, a nuc box is constructed in exactly the same way as a brood box but has only space for 5 frames. With top bar hives the same applies but there is the complication of feeding. A nuc top bar hive has to have the same section as a full top bar hive but also needs the additional space for a feeder. Below is a series of photos of a top bar nuc box.

41. View inside a top bar nuc box showing the sides angled in at 15 degrees. Note the off-centre position of the entrance hole, in the angled side; this places it in the centre of the five drops of comb when in place.

42. Top bar nuc box with 5 top bars to show how drops of comb would be placed from a donor hive.

43. Top bar nuc box with drop-in feeder in place. Note the entrance with small alighting board.

44. Top bar nuc box with crown board. Note that the alighting board doubles up as a hive closure by rotating it.

45. Top bar nuc box with roof, again with entrance closed.

In addition to artificially creating a new colony by means of creating a nuc from drops of comb with queen cells there is the more natural way of letting nature take its course; by this I mean let the bees swarm. Top bar hive beekeeping is by-and-large all about letting the bees do what they want to do. The natural way is to let them swarm and hive the swarm into a new top bar hive or nuc box. From the beekeeper's point of view this may not be the best way, as the possibility of losing a swarm is great. However, I believe that with or without managing to get the swarm, the donor hive always builds up more rapidly than when a number of drops of comb are taken and replaced with new top bars. Similarly a swarm builds up to produce a better new colony.

So, many aspects of top bar hive beekeeping, (once the fundamental problems of inspection and manipulation are overcome), will, as with all methods of beekeeping, settle and become norms and will suit beekeepers as individuals.

The standardisation of a top bar hive is essential, if top bar hive beekeeping is to become part of the beekeeping world. The standardisation need not encompass every aspect of top bar hive construction but must be extensive enough so that would-be top bar beekeepers can purchase items such as feeders, customised top bars, and starter strips of beeswax foundation or some other material. At the time of publication of this booklet I will be working closely with a hive manufacturer so that a "standard" top bar hive can be made available.

Lightning Source UK Ltd.
Milton Keynes UK
UKOW06f0638010913

216306UK00002B/5/P